Easy-to-Build Birdhouses

Mary Twitchell

CONTENTS

Introduction 2
Why Build Birdhouses? 2
Choosing the Right Materials ....
Choosing the Right Design ....
Mounting ....
Nesting Platform for Robins,
  Eastern Phoebes, and Barn Swallows 10
An All-Purpose Nesting Box for Chickadees, Titmice,
  Downy Woodpeckers, and Nuthatches 13
The Wren House 16
Nesting Boxes for Wood Ducks 20
Nesting Boxes for Kestrels and Screech Owls 23
Gourd Nesting Boxes 26
Nesting-Box Maintenance: Predators and Pests ... 27
Attracting Birds to Their New Home 30
The Right Time for Installation 31

Introduction

It hardly matters where you live — on that one special morning in early spring just before daybreak, your hushed neighborhood will suddenly burst forth into a cacophony of singing, tweeting, chirping, cawing, squawking, warbling. The birds are back, and it's time again for their yearly rituals of mating, nesting, and raising a family.

With a little encouragement, some of these avians can become regular seasonal residents (robins, flickers, wrens, sparrows); others (chickadees, titmice, Downy Woodpeckers) may be induced to become year-round companions, providing continual enjoyment for your whole family.

Initially you may learn to identify different species, then their song patterns, nesting habits, and preferred habitats. Soon you will be erecting feeders and setting out birdhouses, or nesting boxes, to encourage birds to feed, court, and breed in your yard.

Why Build Birdhouses?

In cities and in the suburbs where we have converted woodlots, forests, marshy land, and open spaces to housing and asphalt, we have depleted natural bird shelter. Yearly cleanup of brush and dead limbs have reduced the number of sites for nest building. Rows of wooden fence posts, once home to bluebirds and wrens, have now been replaced with metal stakes; and wolf trees formerly left as boundary markers are now mercilessly pruned or removed, especially if they are near power lines. Natural shelter is increasingly difficult for birds to find. Yet attracting birds is beneficial; they love to eat the insects we love to swat. Birds eat many times their weight in insects, keep weeds in check (by eating the seeds), and propagate trees and shrubs by inadvertently dropping seeds in flight.

Of the 650 different species of North American birds, approximately 50 will accept a nesting box in which to raise a family. These species look for a protected, secure home. In their natural habitat,

> **What's in a Name?**
>
> Although *birdhouse* is the traditional name for these structures, the more descriptive term often used by experts in the avian field is *nesting box*. I'll use the terms interchangeably herein.

cavity nesters drill out holes in dead branches or tree trunks (chickadees, titmice, woodpeckers); use holes vacated by other birds (wrens, bluebirds, swallows, flycatchers); take up residence in hollows caused by lightning, fungal infection, or insects; or use hidden tree crotches for nest supports (warblers, robins, finches).

Above all, the fledglings of cavity nesters need protection from predators. This means a small, enclosed nesting area and a small entrance hole. However, one size of nesting box does not fit all. Before choosing which nesting box to build, decide which birds you want to attract. Your local Audubon Society can provide information on the bird species in your area.

Choosing the Right Materials

The materials from which a birdhouse is made play a large part in attracting and safely hosting birds. If you are making your own, starting out with the best of time-tested materials will increase your chances of success; then again, if you are buying a premade birdhouse, you will still want to make sure that it is made of the appropriate materials.

The Best Types of Wood

Wood is the best material for constructing a birdhouse. It is readily available and easy to work with. Cedar, cypress, and redwood last the longest but are expensive and may need to be predrilled before nailing. Fir weathers well, but pine is probably the most available. Lumberyards carry ¾-inch (1.9 cm) stock, which is an ideal thickness; it provides adequate insulation from both heat and cold, is durable enough to resist warping, and is easy to use with hand tools. CDX plywood (exterior grade) of ¾-inch thickness can be substituted. Plywoods made for interior use will quickly delaminate, whereas the layers (plies) of CDX are bonded together with a marine glue that can withstand exposure to different weather conditions.

Rough-cut lumber (before it is planed) and slab ends from a milling operation (sawmill waste with the bark still on) are inexpensive, appropriate, and rustic looking. They may, however, demand more ingenuity and craftsmanship if you're to make the pieces fit.

Gourds are a natural substitute. You can grow them in your garden for next year's supply of nesting boxes (see page 26).

Box Materials to Avoid

Materials to avoid include metals, plastics, coffee cans, bleach bottles, milk cartons, and PVC drainpipe. All of these are very thin and will heat up quickly. Metal roofs are used on Purple Martin houses (to lighten the construction) and sometimes on Wood Duck houses (to protect against raccoons) but should be avoided on boxes for other birds. Also avoid wood treated with preservatives and plywoods made for interior use.

Glue

To increase the life of the nesting box, glue all joints before nailing or screwing. The glue will give a weathertight fit and hold the boards together over time. Use an exterior grade of wood glue; it will be yellow in color and is waterproof. The glue will not bond, however, if any of the surfaces have been painted or varnished.

Fasteners

Galvanized nails will last longer than steel ones; ring-shank nails will hold better than smooth nails. Brass or stainless-steel plasterboard screws are long lived, resist corrosion, and do not stain the wood; brass hinges are preferable to steel ones. Use screws, not nails, when attaching boxes to trees or posts to make seasonal removal easier.

Construction Tip

After nailing the boxes together, check to make sure that none of the nails protrudes into the cavity and that all joints fit tightly. Birds can catch, or even break off, a claw in a crack left between boards.

Paint and Stains

There is no need to paint your birdhouse, but if you wish to, use an exterior, water-based (latex) paint. Never paint the entrance — it makes it difficult for birds to get a good foothold. And never paint or stain the inside of the box, as the chemical fumes are hazardous.

Purple Martin houses, unlike other nesting houses, are very exposed in that they are mounted in open areas 10 feet (3.1 m) off the ground. Thus, their roofs are painted white to reflect the heat. All other birds, however, will avoid a white box; to them it means a more conspicuous home, hence more obvious to predators.

If you decide to paint, use dull, drab colors such as green, gray, tan, and light brown and do the painting in fall so that all vapors will have dissipated by spring. The wood may also be treated with a nontoxic linseed oil.

Choosing the Right Design

Although each species has its own specific requirements, all birdhouses should provide the following:

- Shelter
- Protection from the elements (wind, rain, intense sun)
- Ventilation without drafts
- Insulation
- Drainage
- Durability
- Freedom from chemical fumes (such as pressure-treated lumber, preservatives, stains, and paint, especially when they are applied to the entrance or the interior walls of the birdhouse)

The nesting box should also be accessible for easy cleaning between broods, monitoring of the nest-building process and the fledglings, and inspection (so that you can remove unwelcome tenants like mice, snakes, and House Sparrows).

This bulletin gives five basic house plans. The overall design dimensions (height, length, width), the size of the entrance hole, and the height from the entrance to the floor will vary depending on the species you are trying to attract. Above all, birds like a snug fit. Nesting boxes that are too small cause overcrowding; boxes that are too big make it impossible for the mother to keep the eggs, then chicks, warm. Although the dimensions given with each plan must be followed, you can let your sense of aesthetic embellishment be your guide. Birds don't care whether they are nesting in a French château, a Victorian gingerbread, or an Italian villa.

The Right-Sized Entrance Hole

The entrance hole is drilled near the top of the front panel. Its size and shape will vary according to the bird species; for chickadees it is 1⅛ inches (2.9 cm) in diameter, while for barn owls it is 6 inches (15.2 cm) in diameter. The dimension is based on what is just large enough for the desired species but small enough to exclude predators.

The entrance hole should be at least 6 inches (15.2 cm) above the floor so that the nestlings won't fall out. If the front panel is smooth, it will be difficult for the young to climb up to the entrance hole when the time comes for them to leave. Rough-cut lumber will replicate the rough interior surface of a natural cavity. If you use finished lumber, you'll want to rough up the inside surface from the bottom of the nesting box to the entrance hole to make egress easier for the fledglings — cut multiple grooves ⅛ inch (0.3 cm) deep or tack wooden cleats or wire mesh onto the wall.

Adequate Ventilation

To keep from becoming hotboxes, birdhouses need good ventilation. Air vents provide even temperatures and fresh air; they also prevent the boxes from being too dark. Vents are easy to include by leaving a

---

Protection Begins with the Design

The designs of most nesting boxes are formulated not just to accommodate a particular species of bird, but also to protect that bird from its most common predators. Therefore, it's important to pay attention to the design requirements given for each box.

For example, oversized entrance holes let in birds larger than the desired species — that is, predators. Multiple entrances (except on Purple Martin houses) will permit, even encourage, unwanted visits from House Sparrows and starlings.

In addition, perches are usually not required by the occupant but are handy for the intruder. Perches give House Sparrows, starlings, and other predators a place to wait for lunch. These pests, especially when perched just outside the entrance hole, can cause so much anxiety and disruption to the brooding female that she will abandon the nest.

¼-inch (0.6 cm) gap between the sidewalls and the roof, or by drilling vent holes along the top of each side panel.

Extended Roof

Regardless of design, the birdhouse roof should slope and overhang on the sides. On the front, the roof should extend 2 to 3 inches (5.1–7.6 cm) beyond the box. This extended eave protects the entrance hole from adverse weather (driving rains, wind, sun) and hinders predators from reaching into the nest.

> **Ensuring a Tight Fit**
>
> The sloping roof will fit better if it is beveled along its back edge, where it meets the back. However, if angled cuts are too difficult, nail a ½-inch-diameter (1.3 cm) dowel in the open crack between the back and roof to prevent rain seepage.

To encourage water to shed on finished lumber — especially if the roof is level or almost level — cut a groove 1 inch (2.5 cm) in from the outer edge and ⅛ inch (0.3 cm) deep on the underside of the roof. This groove will prevent rainwater from draining back into the interior of the box.

Wet-Weather Drainage

For the duration of their protected custody, young birds (fledglings) are trapped; not only can they suffocate from excessive heat, but if they become wet, they can die of hypothermia. Any water collecting in the bottom of a birdhouse, which will quickly waterlog the nesting material, is potentially dangerous to the young birds.

A sloping and overhanging roof offers some protection, but to ensure that the nesting box floor will drain properly, cut off the small triangular tip of each corner or drill ¼-inch (0.6 cm) holes in the floor.

Additional measures to increase water protection include:

- Drilling entrance holes on an upward slant
- Recessing box floors ¼ inch (0.6 cm) from the vertical sides, front, and back to let the water drip off
- Nailing a small strip of metal or a piece of roofing paper over the ridgeline to prevent the roof from leaking

Accessibility for Cleaning and Monitoring

For periodic inspections and for cleaning purposes, make the roof or one of the nesting-box walls removable. The removable piece should screw or latch securely — remember, predators are motivated and clever. If you'll be inspecting the nest repeatedly, a removable roof causes the least disturbance.

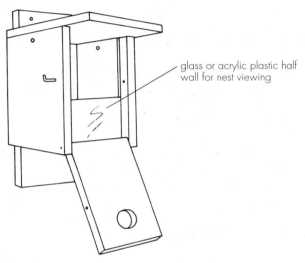

glass or acrylic plastic half wall for nest viewing

The best design includes a detachable roof for inspections and a side panel that pivots for cleaning. A removable sidewall can be fitted with a glass half wall, which will eliminate excessive commotion but also let eager eyes watch the mother's progress.

Monitoring Caution

As baby birds get ready to fledge, frequent inspection becomes risky. Interruptions just before the young are about to fly may encourage the chicks to fledge before they are strong enough.

Mounting

Boxes can be hung from wire; nailed, screwed, or bolted to trees or wooden posts; or mounted on iron pipes and attached with a metal pipe flange. Whichever method you use, know that your box is well constructed and that its inhabitants are safe from predators.

If hanging your nesting box, use flexible 9-gauge wire or vinyl-coated clothesline. Be sure to use an eyebolt, nut, and washer attached to a sturdy roof; a screw eye can loosen and allow the box to fall.

Boxes can be mounted on top of wooden posts with four L-bracket mounts. If mounting your box to a wooden post or tree, be sure to use a predator guard (see pages 27–29), as these fixtures are easy for raccoons, cats, and other predators to climb.

Metal pipe and flange mounts are relatively expensive, but also long lived and very sturdy. If mounting your box with a pipe and flange, do not use any pipe less than ¾" (1.9 cm) in diameter.

Nesting Platform for Robins, Eastern Phoebes, and Barn Swallows

Robins, phoebes, and Barn Swallows will not use an enclosed birdhouse; they prefer a shelter with one or more open sides. The bottom of the shelf platform (bottom) for Eastern Phoebes and Barn Swallows is 6" x 6" inches (15.2 x 15.2 cm); for robins it is 6" x 8 " inches (15.2 x 20.3 cm).

Hang the platform in partial shade, either from the main branch of a tree or under a shed or porch overhang. To help the birds with their nest building, place some clay in a nearby puddle. The birds will use this to cement together the twigs that form their cup-shaped nests.

Robins have adapted well to suburban life and will be easy to lure to backyard nesting platforms.

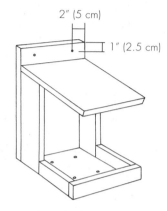

The nesting platform for robins, Eastern Phoebes, and Barn Swallows

Platform Variations

There are many possible variations of the nesting platform design, as can be seen by the examples shown here. In areas protected from the elements, you can even install nesting platforms that are completely open.

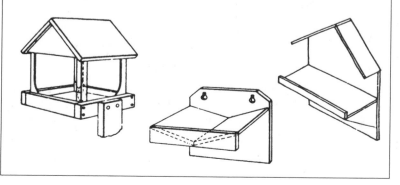

Materials

- ¾" x 7½" x 25" (1.9 x 19.1 x 63.5 cm) wood
- ¾" x 8" x 6" (1.9 x 20.3 x 15.2 cm) wood
- ¾" x 6" x 11½" (1.9 x 15.2 x 29.2 cm) wood
- ¾" x 1½" x 17½" (1.9 x 3.8 x 44.5 cm) wood
- Eighteen 1⅝" (4.1 cm) stainless-steel drywall screws or 6d galvanized ring-shank nails
- Two brass or galvanized #6 x 2" (5.1 cm) wood screws and washers to fit
- Glue

Tools

- Tape measure
- Carpenter's square
- Pencil
- Circular saw or handsaw and miter box
- Hand or power drill and drill bits: ¼" (0.635 cm), ⅜" (0.953 cm)
- Phillips-head screwdriver or power drill fitted with screwdriver bit

Cutting diagram

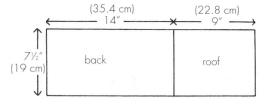
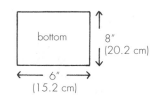
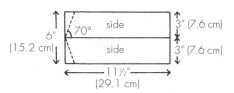
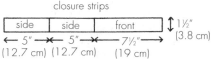

Directions

1. Cut one end of each *side* at a 70° angle.
2. In the *bottom*, drill four ¼" (0.6 cm) drainage holes. Then align each *side* with the back of the *bottom*, glue both faces, and nail or screw together.

Cutting Note

For Barn Swallows and phoebes, the bottom should be cut to 6" x 6" (15.2 x 15.2 cm) and the side closure strips should be measured off at 3" (7.6 cm).

3. If you will be mounting the platform, predrill two ⅜" (1.0 cm) holes 1" (2.5 cm) down from the top and 2" (5.1 cm) in from either side of the *back*. Align the *back* with the *sides* and *bottom*; glue and screw or nail together.
4. Bevel-cut the *roof* at 70 degrees along one of the 7½" (19.1 cm) edges. Glue and screw or nail the *roof* to the *sides* and the *back* into the *roof*.
5. Glue and screw or nail 1 x 2 inch (2.5 x 5.1 cm) front and side *closure strips* around the edge of the *bottom*.

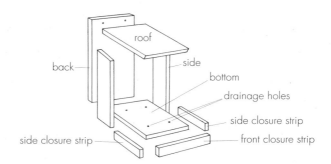

Mounting

If you're going to attach the platform to a tree, building, or post, secure it through the *back* with 2" (5.1 cm) screws and washers.

For robins: Mount a robin platform 6' to 15' (1.8–4.6 m) above the ground.

For phoebes and Barn Swallows: Mount platforms for phoebes and Barn Swallows 8' to 12' (2.4–3.7 m) above the ground.

Appropriately enough, Barn Swallows often find the multitude of open platforms in barns quite inviting.

An All-Purpose Nesting Box for Chickadees, Titmice, Downy Woodpeckers, and Nuthatches

Chickadees, titmice, Downy Woodpeckers, and nuthatches prefer the same habitat — the borders of woodlands and woodland clearings. They also prefer many of the same foods, including insects, nutmeats, and suet. If the nesting box is placed near some shrubbery, which provides good cover, and feeders are set up close by, these birds are even more likely to use a nesting box.

Erect the nesting box at eye level, either secured to a tree trunk or hung from a tree limb. Place 2 to 3 inches (5.1–7.6 cm) of sawdust or wood chips in the bottom for nest building.

Materials

¾" x 6½" x 7" (1.9 x 16.5 x 17.8 cm) wood
¾" x 5½" x 25" (1.9 x 14.0 x 63.5 cm) wood
¾" x 4" x 22½" (1.9 x 10.2 x 57.2 cm) wood
One ½" x 5½" (1.3 x 14.0 cm) hardwood dowel
Sixteen 1⅝" (4.1 cm) stainless-steel drywall screws or 6d galvanized ring-shank nails
Two 6d galvanized finishing nails
Three 2d galvanized finishing nails
Six brass or galvanized #6 x 2" (5.1 cm) wood screws and washers to fit ⅝" (1.6 cm) staples or wire brads
3" x 6" (7.6 x 15.2 cm) piece of ¼" (0.635 cm) galvanized wire mesh (hardware cloth)
One L-shaped screw (optional)
Glue

Tools

Tape measure
Carpenter's square
Pencil
Jigsaw
Circular saw or handsaw and miter box
Hand or power drill and drill bits: ¹⁄₁₆" (0.159 cm), ⅛" (0.318 cm), ¼" (0.635 cm)
1⅛" (2.858 cm) paddle bit, hole saw, or expansion bit
Phillips-head screwdriver or power drill fitted with screwdriver bit
Hammer
Sandpaper (optional)
Staple gun (optional)

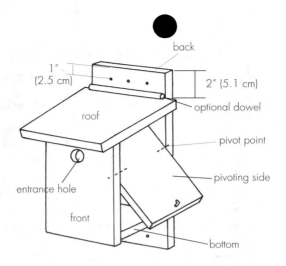

The all-purpose nesting box

Cutting diagram

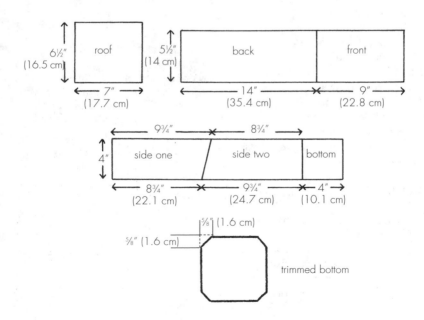

Directions

1. Cut off a triangle ⅝" x ⅝" (1.6 x 1.6 cm) from each corner of the *bottom* to provide drainage holes.
2. Glue and screw or nail *side one* to the *bottom*, recessing the bottom ¼" (0.6 cm) to create a drip edge.
3. On the *front*, mark the center of the entrance hole by measuring down 1⅝" (4.1 cm) from the midpoint of the top of the *front*. Drill the 1⅛" (2.9 cm) hole and gently sand any rough edges. Tack or staple the hardware cloth inside the box and below the entrance hole. Glue and screw or nail the *front* to *side one*, aligning their bottom edges. Glue and screw or nail the *front* to the *bottom*.
4. To allow for mounting, predrill three ¼" (0.6 cm) holes spaced 1" (2.5 cm) apart and along a line 1" from the top and 1" from the bottom of the *back*. Then measure up 2" (5.1 cm) from the bottom of the *back* and draw a line. This marks the bottom edge of *side one*. Align the *side* with this line and glue and nail or screw the *back* to the *bottom* and to *side one*.
5. To attach the roof, make a 76° angled cut along one of the 6½" (16.5 cm) edges of the *roof*. Position the *roof* so that there will be a ½" (1.3 cm) overhang on *side one* and a 1½" (3.8 cm) overhang in front. Glue and screw or nail the *roof* to the *front* and through the *back* into the *roof*. To ensure that the roof fits tightly, predrill three holes in the dowel with a ¹⁄₁₆" (0.159 cm) bit and nail above the *roof* and into the *back* with 2d nails.
6. *Side two* will pivot for easy cleaning of the nesting box. From the bottom of *side two*, measure up 6" (15.2 cm). Draw a line. Set *side two* in place on the nesting box, remembering to leave the ¼" (0.6 cm) vent gap. With a ⅛" (0.318 cm) bit, drill through the *front* into *side two*, then through the *back* into *side two*. Drive a 6d nail into each set of holes. The two finishing nails will allow *side two* to pivot. The pivot nails must be exactly opposite each other for the side to open easily. To secure *side two*, screw into the *bottom* or use an L-shaped screw.

Mounting

Mount the box on a tree or post with 2" (5.1 cm) wood screws and washers.

The Wren House

Wrens are the least fussy of the cavity-nesting birds. They will use almost any accommodation as long as it's small and has an elliptical slot, not a circular entrance hole. Because wrens fill their nests with a tangled mess of twigs, they need the larger elliptical shape for maneuvering sticks into the box.

The male works on a number of nests until he finally builds one to the female's liking. Hanging multiple boxes will ensure that the nest of ultimate choice is also in your yard. Wrens are quite sociable; even if the boxes are close to your house, wrens won't be reluctant about using them.

The wren house

Experiment with different designs and materials for wren boxes. Any rustic design (that is, made from rough-cut lumber) will do. If you are using finished wood, however, the young will have difficulty negotiating the smooth surface to reach the entrance hole. To assist them in climbing out, roughen up the inside of the front piece. Cut ⅛-inch (0.3 cm) saw kerfs, add wooden cleats, or tack hardware cloth below the entrance hole to give the young something to climb.

Materials

- ¾" x 6" x 30" (1.9 x 15.2 x 76.2 cm) wood
- ¾" x 5½" x 11" (1.9 x 14.0 x 28.0 cm) wood
- ¾" x 4" x 4" (1.9 x 10.2 x 10.2 cm) wood
- Sixteen 1⅝" (4.1 cm) stainless-steel plasterboard screws or 6d galvanized ring-shank nails
- 2" x 4" (5.1 x 10.2 cm) piece of ¼" (0.635 cm) galvanized wire mesh (hardware cloth)
- ⅝" (1.0 cm) staples or wire brads
- One eyebolt, nut, and washer (optional, depending on whether the wren house will be hung)
- Glue

Tools

- Tape measure
- Carpenter's square
- Pencil
- Jigsaw
- Circular saw or handsaw and miter box
- Hand or power drill and ⅜" (0.953 cm) drill bit
- Phillips-head screwdriver or power drill fitted with screwdriver bit
- Hammer
- Sandpaper (optional)
- Staple gun (optional)

Cutting diagram

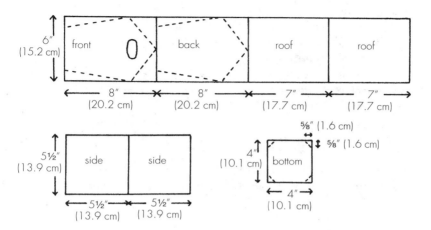

Directions

1. Cut a ⅝" x ⅝" (1.6 x 1.6 cm) triangle off each corner of the *bottom* to create drainage holes.
2. To mark cutting lines on the *front* and *back*, draw a centerline (AB) the length of each board, 3" (7.6 cm) in along the 6" (15.2 cm) edge. At the bottom of each board, measure 2" (5.1 cm) to either side of the centerline and mark points C and D. At point A, mark for the roof cuts, measuring 50° in either direction of the centerline from point A to points E and F. Draw lines AE and AF. Cut along lines AE, AF, CE, and DF.

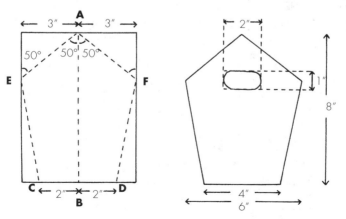

3. On the *front*, draw a line between points E and F. Where this line intersects the centerline (G) is the center of the 1" × 2" (2.5 × 5.1 cm) entrance hole. From point G, measure 1" to either side along line EF and mark. Then from point G, measure ½" (1.3 cm) to either side along the centerline (AB). Connect the points to form an oblong entrance opening.

    To cut out the entrance hole, drill ⅜" (1.0 cm) holes at each of the four points you have marked. Be sure the drill bit removes wood only within the oblong. Connect the holes using a jigsaw. Use a rasp or sandpaper to smooth any rough edges.
4. Tack or staple hardware cloth inside the box and below the entrance hole.
5. Bevel-cut each *side* at 50° along one of the 5½" (14.0 cm) edges.
6. Bevel-cut the *roof* at 50° along one of the 7" (17.8 cm) edges.

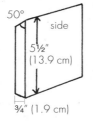 

7. Glue and screw or nail the *front* to the *bottom*, recessing the *bottom* ¼" (0.6 cm). Glue and screw or nail the *sides* to the *bottom* and *front*. The *sides* should be flush with the *front* and extend ¼" beyond the *bottom*.
8. The beveled edges of the *roof* should create a tight fit along the ridge. Glue and screw or nail one side of the *roof* to the *sides, front,* and *back*. The *roof* should be flush with the *back*; the overhang should be along the *sides* and *front*. The second section should be screwed *only* so that it can be easily detached for inspections and for cleaning out the nest cavity.

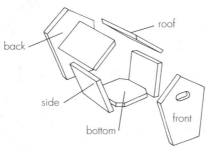

Mounting

Screw an eyebolt into the *roof* if the nesting box will be hung. Hang the box at eye level on a partially sunlit tree limb.

Wren Housing

Wrens are more likely to take advantage of quarters you provide for them than any other bird. Usually, because of competition and territorial prerogatives, it's impossible to attract more than one nesting pair of any given species to a normal-sized lot. Not so with wrens. Some males will mate with more than one female, and they don't seem to trouble themselves with separating the families to avoid scandal. With their cheerful songs, amusing antics, and appetite for insects, wrens are welcome tenants.

Carolina Wren

Of all the cavity-nesting birds that will use birdhouses, wrens are the most likely to seek peculiar nesting locations. They've been known to use discarded shoes, tin cans, and the leg of a child's blue jeans hung up to dry.

A Carolina Wren couple built a nest in the middle of my brother and sister-in-law's brand-new double pink vining geranium. Hardly had Bill placed the plant in a hanging basket attached to the overhang just above their front steps when wrens set up housekeeping. Mama laid her six pinkish brown spotted eggs smack in the middle of the geranium. Five of the eggs hatched. While the babies were still nestlings, Dee watered the plant very carefully, almost by droplets, in the hope of preserving it without damaging the birds. Four of the five nestlings survived to fledgling size and departed. When they left the nest, Dee investigated the geranium and found that the deep cuplike structure had displaced most of the soil in the middle of the plant. She replaced it, and the plant — though scarcely robust — survived.

— Jan Mahnken,
The Backyard Bird-Lover's Guide
(Storey Publishing, 1998)

Nesting Boxes for Wood Ducks

Wood Ducks prefer nesting in tree cavities near water and surrounded by plenty of cover. Place the boxes in or over the water of swamps, lakes, or ponds, or set back 30 to 100 feet (9.2–30.5 m) from shore — boxes directly on the water's edge are more prone to raccoon predation. If water is within a mile (1.6 km) of the box, Wood Ducks will still use the boxes for nesting; the mother will take her young to water, but along this trek the ducklings are extremely vulnerable.

Mount nesting boxes on a wood or metal post. To further protect the fledglings from predators, add a cone baffle (see page 28) 3 feet (0.9 m) above the level of the water or ground. Multiple boxes may be mounted on the same post.

If the boxes are mounted in trees, metal guards are even more imperative to protect the young from predators, especially raccoons.

Ducks do not collect nest materials; line the box with 3 inches (7.6 cm) of wood shavings or sawdust prior to each nesting season.

Materials

- ¾" x 12" x 58½" (1.9 x 30.5 x 148.6 cm) wood
- ¾" x 13½" x 51" (1.9 x 34.3 x 129.5 cm) wood
- ¾" x 15½" x 15½" (1.9 x 39.4 x 39.4 cm) wood
- Two 4" lag screws and washers
- One brass hinge ½" x 3" (1.3 x 7.6 cm) and screws
- Thirty 1⅝" stainless-steel drywall screws or 6d galvanized ring-shank nails
- One 4" x 16" (10.2 x 40.6 cm) piece of ¼" (0.635 cm) galvanized wire mesh (hardware cloth)
- ⅝" (1.6 cm) staples or wire brads

Tools

- Tape measure
- Carpenter's square
- Pencil
- Jigsaw
- Circular saw or handsaw and miter box
- Hand or power drill and drill bits: ¼" (0.635 cm), ⅜" (0.953 cm)
- Phillips-head screwdriver or power drill fitted with screwdriver bit
- Hammer
- Adjustable wrench
- Sandpaper (optional)
- Staple gun (optional)

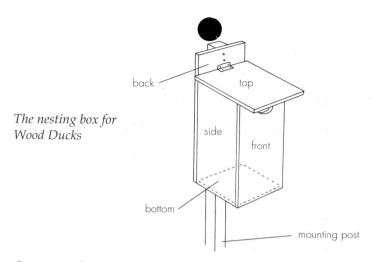

The nesting box for Wood Ducks

Cutting diagram

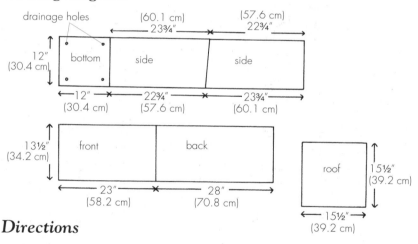

Directions

1. Drill four holes in the *bottom*, ¼" (0.6 cm) in diameter, for drainage.
2. Glue and screw or nail the *sides* to the *bottom* with the *side* slants sloping in the same direction. Recess the *bottom* ¼" (0.6 cm) on each side before attaching.
3. Align the *back* with the bottom of the *sides* and glue and screw or nail the *back* to the *sides* and *bottom*. The *back* will extend above the *sides* for mounting the nesting box.
4. Bevel-cut (85°) along one edge of the *roof*. Place the *roof* on the box so that it aligns with the *back*, overhangs the *sides* by 1" (2.5 cm), ands overhangs the *front*. Center the hinge and screw it into the *roof* and the *back*.

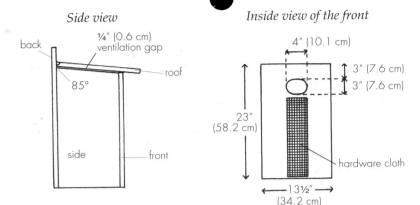

5. To make the entrance hole on the *front*, mark the center of the *front* along the top edge. From this point, draw a centerline placing marks at 3" (7.6 cm) and 6" (15.2 cm). These represent the top and bottom of the entrance hole.

   Again on the top edge of the *front*, measure 2" (5.1 cm) to either side of the center and mark. Draw two lines at 90°. Along each of these lines, place a mark at 4½" (11.4 cm). These marks represent the widest points of the entrance hole. Connect these points to form an oblong entrance opening.

   Drill ⅜" (1.0 cm) holes at each of the four points you have marked. Be sure the drill bit removes wood only within the oblong. To remove the rest of the wood, connect the holes with a jigsaw. Use a rasp or sandpaper to smooth any rough edges.

6. Tack or staple a strip of mesh hardware cloth to the inside of the *front* and below the entrance hole. This will assist the fledglings when it's time for them to leave the box. Be sure there are no sharp edges. Glue and screw or nail the *front* to the *sides* and *bottom*.

Mounting

If the Wood Duck nesting box is to be mounted on a 4" x 4" (10.2 x 10.2 cm) post, predrill two ⅜" (1.0 cm) holes along the centerline of the *back*. Erect the 4" x 4" cedar post. Mount the box 10' (3.1 m) above the ground or 5' to 6' (1.5–1.8 m) above water. Slip the washers over the lag screws and lag-screw into the mounting post using an adjustable wrench.

Place 3 to 4 inches (7.6–10.2 cm) of sawdust or wood chips in the bottom of the nesting box for a base.

Nesting Boxes for Kestrels and Screech Owls

Screech Owls and kestrels use nesting boxes of the same dimensions. In fact, if you clean out the nesting box in late spring after the Screech Owl has raised its young, kestrels may take up residence.

Owls prefer to nest in tree cavities or in abandoned woodpecker holes at the edge of a field or in a neglected orchard. Since they seldom build their own nests, they take readily to nesting boxes.

The boxes should be lined with 1 to 2 inches (2.5–5.1 cm) of wood shavings and mounted on a tree trunk in a wooded grove 15 feet (4.6 m) above the ground.

The nesting box for kestrels and Screech Owls

Materials

¾" x 9½" x 31" (1.9 x 24.1 x 78.7 cm) wood
¾" x 8" x 39" (1.9 x 20.3 x 99.1 cm) wood
¾" x 12½" x 11½" (1.9 x 31.8 x 29.2 cm) wood
One ½" hardwood dowel (optional)
Glue
Twenty-six 1⅝" (4.1 cm) stainless-steel drywall screws or 6d galvanized ring-shank nails
Three L-shaped screw hooks
⅝" (1.6 cm) staples or wire brads
4" x 10" (10.2 x 25.4 cm) piece of ¼" (0.635 cm) galvanized wire mesh (hardware cloth)

Tools

Tape measure
Carpenter's square
Pencil
Jigsaw
Circular saw or handsaw and miter box
Hand or power drill and drill bits: ⅛" (0.318 cm), ¼" (0.635 cm), ⅜" (0.953 cm)
3" (7.6 cm) hole saw or expansion bit
Phillips-head screwdriver or power drill fitted with screwdriver bit
Hammer
Sandpaper (optional)
Staple gun (optional)

Cutting diagram

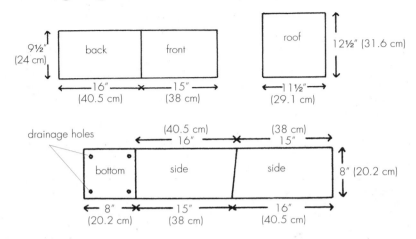

Directions

1. Drill four ¼" (0.6 cm) drainage holes in the *bottom*.
2. For ventilation, drill three holes ⅜" (1.0 cm) in diameter near the top of the *sides*. Glue and screw or nail the *sides* to either side of the *bottom* with the *side* slants sloping in the same direction. Recess the *bottom* ¼" (0.6 cm) on each *side* before attaching.
3. Align the *back* with the *sides*, and glue and screw or nail together.

An exploded view

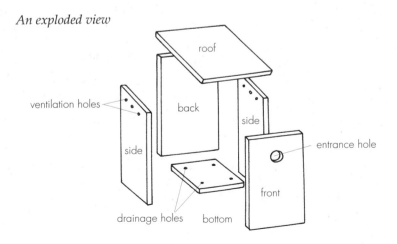

4. The *roof* will align with the back of the box and have a front overhang and a 1" (2.5 cm) overhang on either side. This overhang will assist in shedding water. Glue and screw or nail the *roof* to the *back* and *sides*.
5. To cut the entrance hole on the *front*, mark the center of the *front* along the top edge. From this point, draw a centerline. Measure down along the centerline 3½" (8.9 cm) to mark the center of the 3" (7.6 cm) entrance hole. The center of the circle should be 4¾" (12.1 cm) in from either edge. Cut the entrance hole. Sand the opening to remove any rough edges. Staple or tack the hardware cloth below the entrance hole. Be sure there are no sharp edges.
6. From the bottom of the *front*, measure up 9" (22.9 cm) on either side and draw a line between the points. Hold the *front* in position so it aligns with the *sides*. Along the line you have just drawn, drill through the *sides* into the *front* with an ⅛" (0.318 cm) bit.

Using these pilot holes, screw L-shaped screws through the *sides* into the *front*. The *front* should pivot. To secure the *front*, predrill a third hole through the *roof* into the *front* and screw it in with an L-shaped screw. If you turn each of these screws a quarter turn, the *front* will be easily removable; if you unscrew only the top L-shaped screw, the *front* will pivot for cleaning.

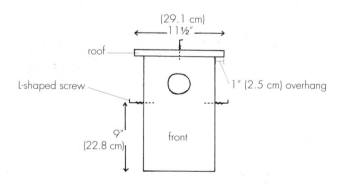

Mounting

For kestrels: Mount the box at least 10' (3.1 m) off the ground.
For Screech Owls: Mount the box 15' (4.6 m) off the ground.

Gourd Nesting Boxes

Many species will take readily to nesting in gourds. Gourds can be grown like pumpkins — simply plant them in your garden in hills of four or five seeds with the hills 4 to 6 feet (1.2–1.8 m) apart. When fully ripe, harvest the gourds, wash them thoroughly with a disinfectant, and spread them on newspapers to dry. While drying (three to four weeks), turn the gourds regularly and spread them out so that they don't touch one another.

After the gourds have dried, cut a hole to the dimensions required by the desired species. Gourds can be tough — you may need to use a keyhole saw or expansion bit to do this. Then use a serrated knife to break up and remove the hard pith and seeds from the inside.

To shield the entrance hole from the sun and rain, you may wish to add a small canopy made from flexible metal or plastic above it. If so, attach the canopy with silicone caulking and let it dry thoroughly before setting outside.

Attach a fastener to the top of the gourd and hang it from a porch roof overhang, tree limb, or pole.

A gourd nesting box with metal canopy

Just the Big Ones

Check the depth from the inside bottom of the gourd to the entrance opening. To give birds enough space to make a nest, try to use gourds that are large enough so that there is at least 5½ inches (14.0 cm) of space from the bottom of the doorway to the bottom of the gourd.

Nesting-Box Maintenance: Predators and Pests

Once you put out bird feeders or nesting boxes, predators become a perpetual concern. Even Fluffy, the once adorable house cat, is turned into terrible Tom and sits crouched and motionless in the grass, soundlessly switching her tail and waiting for the perfect moment to pounce. And clever folks (raccoons and opossums) stake out your yard by night waiting to hector the parents and capture the kids.

Squirrels, mice, or snakes may even have taken up residence before the birds of your choice arrive. Insects and parasites may attempt to cohabit, and there is always the annoyance of sparrows and starlings flitting around trying to distract the mother.

> **Evictions**
>
> Should House Sparrows or European Starlings begin nest building in your box, destroy the nest at once; it's perfectly legal. And evict any unwanted rodents.

To give your birds their best chance, carefully build to each box's specified dimensions (depth, height, distance to entrance hole, and entrance hole size and configuration). These dimensions not only meet the birds' needs but also minimize the ability of predators to wreak havoc.

For starters, try smearing the mounting pole with Vaseline or a mixture of grease and hot red cayenne pepper. If predators continue to harass the nesting birds, however, you may need to implement some more rigorous measures to safeguard your birds.

Squirrel-Proofing

Teeth and gnawing marks around a nesting-box entrance are signs of squirrels attempting to enlarge its hole. Predator guards made of metal will discourage their activities. Cut the metal 1" (2.5 cm) larger than the opening. Remove a center hole equal in size and shape to the entrance hole and tack the guard in place. Be sure there are no burrs or sharp edges along the metal.

Noel Wire Raccoon Guard

The angled roof and overhang of each nesting box make it difficult for predators to reach into the entrance hole from the roof. However, house cats, squirrels, and — especially if your nest is located near swampy water — raccoons and opossums still may be a problem. They will insert a paw and pull out the young or the eggs. Wood shavings on the ground or around the entrance hole are signs of their activity. To discourage these menaces, use a sharp-edged wire guard that makes for a long, uncomfortable reach.

Noel wire raccoon guard

Cone Baffle

If smearing the pole with Vaseline or a mixture of grease and hot red cayenne pepper doesn't work, build a 3-foot (0.9 m) metal cone guard out of 26-gauge sheet metal. The cone should stay in place, but wobble enough to prevent the predator from reaching the nesting box.

To mount a cone on a wooden post: Cut four blocks to 3" x 3" (7.6 x 7.6 cm) and nail to the post just below the nesting box; the blocks will hold the cone guard in place.

To mount a cone on a metal pole: Fit the pole with a hose clamp, a piece of twisted wire, or a wooden block to support the cone.

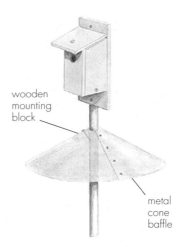

Use wooden blocks to support cone baffles on mounting poles.

Metal Sleeves and Poles

If you know you have a flying Wallenda for a house cat, mount the nesting box on a metal pole. Such poles are the most difficult for predators to climb. Other pests (snakes, for example) and particularly persistent cats and squirrels may be more successfully thwarted if the metal support pole is layered with grease and sprinkled with red cayenne pepper.

Sheet-metal guards on a wooden post or tree where the nesting box is mounted offer additional protection and are best installed before the birds begin nesting. Bend an 18-inch-wide (45.7 cm) sheet tightly around the wooden post or tree and tack or nail it in place. Narrower strips of sheet metal may not be sufficient to defeat a springing cat.

Long metal sleeves will discourage most climbing predators, such as cats, raccoons, and snakes, from disturbing the occupants of nesting boxes.

Leashing Your Pets

Dogs and cats should be leashed or kept inside during nesting time, especially when there are young birds. Bells on collars don't stop cats from being successful hunters.

Attracting Birds to Their New Home

Birds need protective shelter, food, and access to ample water. Make your best attempt to locate nesting boxes in habitat most resembling the bird's natural environment, keeping in mind that some birds, such as Wood Ducks and owls, are more discriminating than others, such as robins, wrens, and chickadees. It may take a year before the birds discover your nesting box, but if it takes longer, reexamine the box and the site. The location you've chosen may not be right for the birds, or perhaps you need to provide some more tempting attractions.

- **Food and water.** Many times birds can be encouraged to nest when the box is close to fruit-bearing shrubs or a birdbath. However, avoid placing boxes right next to a bird feeder.
- **Sheltering foliage.** Check the vegetation around the box. If there is none, there will be nowhere for the mother to shelter the fledglings; if there is too much vegetation, it provides plenty of protection for predators. Or parents-to-be may have already discovered a predator nearby and opted for a safer location.
- **Protected exposure.** Check also that the entrance hole faces south, southwest, or west, and is situated with a clear flight path to its entrance. Do not point the entrance hole in the direction of prevailing winds or storms, or in the path of direct exposure from the afternoon sun. Houses must not overheat. If your summers are extremely hot, point the entrance to the north or east.
- **The right-sized entrance hole.** Ensure that the entrance is the proper size for the species you are trying to entice.

Many small cavity-nesting birds, such as chickadees, titmice, and sparrows, will appreciate good cover near their nesting box.

Nesting Materials

To make the nest-building process easier (and possibly make your birdhouse more attractive to potential nest builders), you can offer birds nesting materials. Robins and Barn Swallows need mud. Songbirds use twigs, feathers, straw, grasses, and leaves to build their nests but will be thankful for short threads, string, yarn, tissues, hair (human or horse), lint from the clothes dryer, and bits of cloth. Long threads or wads of cotton may be more dangerous than useful; they can become tangled in a bird's claws. Add wood shavings to the boxes of chickadees and woodpeckers; they'll use the shavings to shape their own nesting cavities.

Look for other nesting facilities in your immediate locale. Your birds may have found safer sites. There also may be too many nesting boxes in the area. Four small boxes (or one large box) per species per acre is the maximum, and more than one box per tree is too much unless the boxes are for different bird species. Birds are territorial, and competition is worst among those of the same species.

Check also that an unwanted occupant (such as a snake, mouse, or squirrel) isn't already nesting in your box.

The Right Time for Installation

Nesting boxes frequently go unused because they are put out too late in the season. You must install nesting boxes before the migratory birds arrive and breeding begins; dates will vary by region, but generally this means January in the South and mid-March in the North.

Setting out new boxes in late fall is ideal. By spring they will have weathered and become more attractive to birds, which might find a brand-new shelter suspicious. This schedule also gives birds that winter over (woodpeckers, swallows, and nuthatches) time to get used to the boxes; they may even roost in them.

Installing boxes in fall has a further advantage: If you will be mounting the box on a pole or post, the ground may still be frozen in early spring. Driving in posts in late fall will be much easier.

Other Storey Titles You Will Enjoy

The Backyard Birdhouse Book,
by René and Christyna M. Laubach.
A beautifully illustrated guide to identifying,
attracting, and housing some of North America's
most fascinating cavity nesters.
216 pages. Paper. ISBN 978-1-58017-104-5.

The Backyard Bird-Lover's Guide, by Jan Mahnken.
An information-packed reference to understanding
more than 135 species of birds.
320 pages. Paper. ISBN 978-0-88266-927-4.

Birdhouses, by Mark Ramuz & Frank Delicata
A collection of unique projects for
all levels of woodworkers.
128 pages. Paper. ISBN 978-0-88266-917-5.

*Drawn to Nature Through the Journals of
Clare Walker Leslie,* by Clare Walker Leslie.
An invitation to readers to take a moment to slow
down and see nature with renewed appreciation.
176 pages. Paper with flaps. ISBN 978-1-58017-614-9.

The Family Butterfly Book, by Rick Mikula.
Projects, activities, and profiles to celebrate
40 favorite North American species.
176 pages. Paper. ISBN 978-1-58017-292-9.
Hardcover. ISBN 978-1-58017-335-3.

These and other books from Storey Publishing are available
wherever quality books are sold or by calling 1-800-441-5700.
Visit us at *www.storey.com*.